Volume Four

...H
LIN... ...ORIES

Ian L. Wright

London & North Eastern

First published February 1986

Designed by Nigel Trevena
Typeset by Delta Graphics, Falmouth, Cornwall
Halftone photo reproduction by Oxford Litho Plates Ltd, Botley, Oxford
Printed by Century Litho, Penryn, Cornwall
Bound by Booth Bookbinders, Penryn, Cornwall

ISBN 0 906899 19 2

Published by
ATLANTIC TRANSPORT PUBLISHERS
Chough House River Street Truro
Cornwall TR1 2SJ England

Atlantic

TITLE PAGE: The LNER branch line atmosphere is nicely captured in this 1950 view of GN C12 4-4-2T No.7374 at the head of a train of GE and Gresley coaches working on the ex GN Alexandra Palace branch in suburban north London.
Photo: P. M. Alexander MHL

ABOVE: GE J15 about to leave Long Melford with the 3.15pm train for Bury St Edmunds on 6 July 1957.
Photo: Ian L. Wright

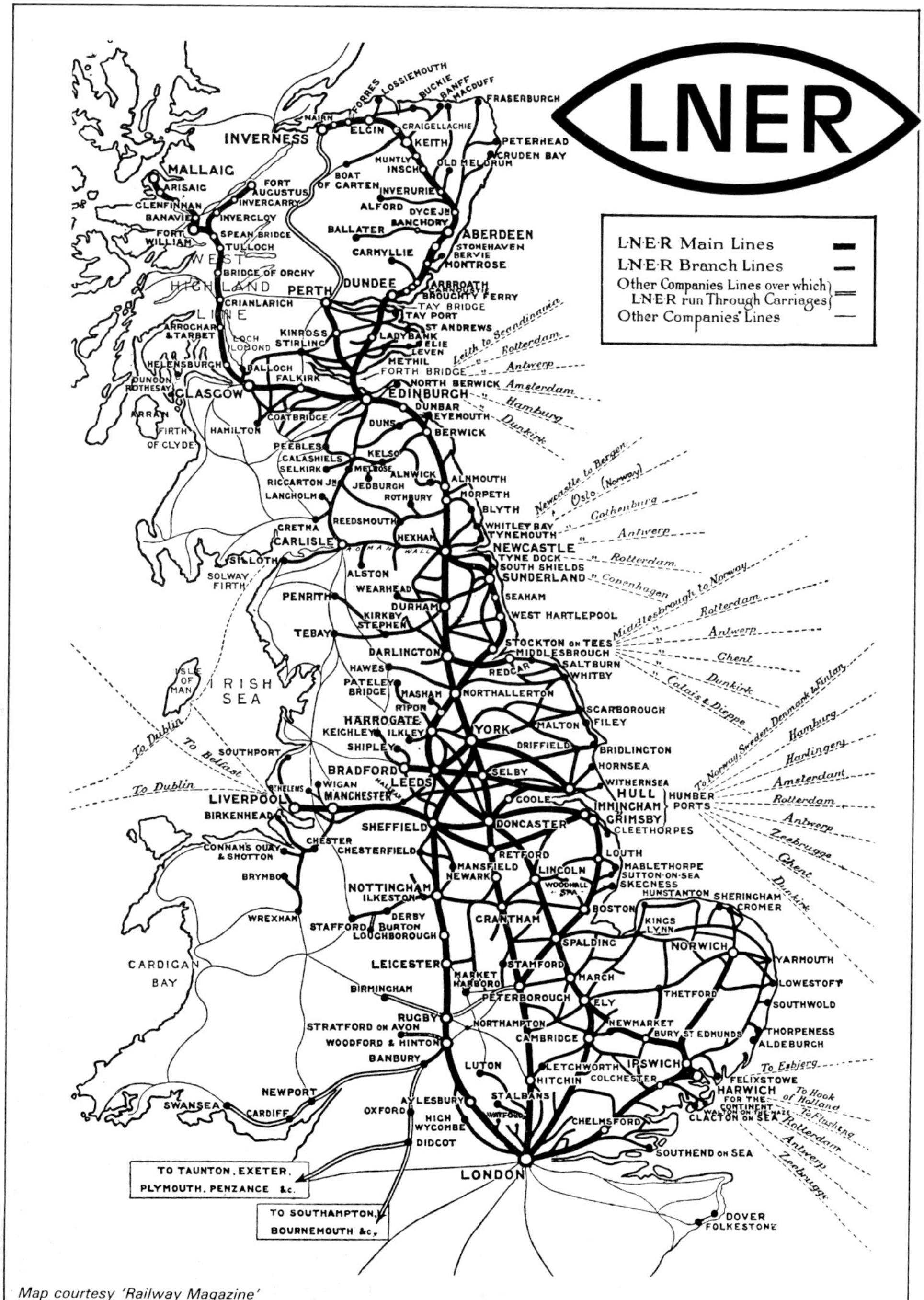

Map courtesy 'Railway Magazine'

FRONT COVER: Sutton Bridge, M&GN, on 28 February 1959. J6 No.64172 waits with a Spalding-South Lynn train.

BACK COVER: Middlesbrough-Whitby and Whitby-Middlesbrough trains reversing at Battersby on 3 May 1958. *Photos: Ian L. Wright*

To my Father
Who loved the North Eastern Railway

The Author

IAN L. WRIGHT NDD ATD runs the Art Department of a Northamptonshire comprehensive school. Ian's father, Ernest Nicholas Wright, a South Shields man, grew up within earshot of the North Eastern Railway on Tyneside and loved its trains. He passed on his love of railways to Ian and in his first home in Barry, Glamorgan, Ernest set up an O gauge model railway. Ian's childhood acquaintance with the LNER was renewed in wartime service in HM Forces. He has explored widely over the national network on lines long since closed, but he describes himself as no more than an intermittent railway photographer. In his life, railway photography has had to be shared with recording canals, trams and architecture.

Acknowledgements

Those of us who remember the branch trains of our youth took them very much for granted. In a very special way they were part of the life of the community, and so familiar that we imagined they would always be there. The train took us into the nearest big town. It was there to take us to school and on holiday to the seaside. During the 1939-1945 war there were restrictions on using a camera to photograph branch lines, but in the twenty years that followed, many of us hurried half way across England to record the charm and character of their diminishing number before it was too late.

I am grateful for the help and encouragement of Nigel Trevena and Richard Riley. The photographic collections of Richard Riley and others, generously put at my disposal, are gratefully acknowledged. Making a selection of photographs that will please everybody is the hardest task of all — so large a system to represent in one modest book — but a geographical survey has been chosen as the best way to present the theme. Ian L. Wright

The LNER Branch Line

The London & North Eastern Railway, formed in 1923, embraced much of England and Scotland. South to north, it stretched from Kings Cross to Lossiemouth, a distance of over 600 miles, though a Great Eastern or North British enthusiast could argue that it ran from North Woolwich to Mallaig! The LNER had East Anglia almost to itself, and most of Yorkshire. There were branches that crossed the hostile Pennines and others wandering into the Cheviot border country where it was hard for a railway to make a living. Through the enterprise of that great penetrator, the Great Central Railway, the LNER even owned lines in Wales.

Perhaps there was only one feature that unified the LNER branch train: its livery. Varnished teak and beautiful gold and shaded transfer numerals for the coaches and apple green or black for the engine. The rest was a mirror to pre-grouping practice — an astonishingly diverse collection of engines and carriage stock that had been turned out of Doncaster, Darlington, Stratford and Cowlairs up to forty years before. Time seemed to have stood still. A North Eastern G5 might haunt a Yorkshire branch line for decades, right through the Grouping and well on after Nationalisation. The 1930s brought economies and a sad list of closures as country passengers left the railway to use the more convenient buses. BR closures accelerated as more people bought cars. Diesel railcars achieved some operating economies from 1954 and prolonged the life of some branch lines, but the end of an era came in the 1960s with Dr Beeching's axeing of nearly 2,000 stations and the removal of the stopping passenger train. The LNER branch train may have been experienced with inconvenience but it is remembered with nostalgia. In this affectionate look back at a world that is gone we extend our thanks to the photographers who had the vision to capture these images of the trains they knew.

ABOVE: A memory of the Stour Valley line. The 1.22pm SO Cambridge-Colchester train headed by GE 2-4-0 No.62789 calls at Haverhill on 6 July 1957. In the background is a J15 ready to follow with a Colne Valley & Halstead line train.
Photo: Ian L. Wright

IN RURAL ESSEX

ABOVE: In 1949, London Transport's Central Line electric trains replaced GE line steam trains on the Epping service, but from Epping to Ongar a steam push-pull train was kept on. The LT electric trains finally took over from steam in 1957. GE F5 2-4-2T No.67203 disturbs the peace of Epping Forest with an afternoon push-pull working from Ongar to Epping on 18 May 1957. *Photo: R. C. Riley*

RIGHT: GE G4 0-4-4T No.8109 heads a train of vintage GE six wheelers at Woodham Ferrers on the cross country branch to Maldon East in April 1932. A week after the outbreak of the Second World War this poorly patronised passenger service had gone. *Photo: Dr Ian C. Allen*

LONDON BRANCHES — GE

Great Eastern Jazz trains: Richard Riley's spirited evocation of the Liverpool Street Jazz trains is a far cry from the many rural branch line scenes that predominate in this book. It is essential though, to remember that the GE ran the most intensive steam suburban services in the world out of Liverpool Street, calling for smart locomotive work. Quick turnround and rapid acceleration were achieved with Holden's small wheeled 0-6-0 tanks on the trains to Ilford, Enfield and Chingford. Later, from 1915, the smart N7s of Alfred Hill took over much of the Jazz train work. In BR days GE N7 0-6-2T No.69665 is seen blasting out of Liverpool Street with a train of Gresley articulated stock bound for Enfield Town on 23 September 1958.

The name "Jazz Train" originated in 1920 when station layouts were modified to enable the service to be intensified — different coloured lines above the doors indicating 1st and 2nd class compartments. A similar idea has been adopted in recent years by BR. *Photo: R. C. Riley*

RIGHT: A number of the appealing little GE 2-4-2 tanks (LNER F7 class) were motor fitted for push-pull working on the GER. Part of the charm of these engines was the enormous roomy and well-glazed cab contrasting with the small dimensions of the boiler. It was no wonder the enginemen called them "Crystal Palaces". In this picture, taken at Palace Gates in 1937, No.8305 prepares to leave on the shuttle service to Seven Sisters. *Photo: R. C. Riley Collection*

LONDON BRANCHES — GN

BELOW: The GNR operated three steam branches to the Northern Heights, in suburban North London, over steep gradients westward from a junction at Finsbury Park. These were the branches to Edgware, High Barnet and Alexandra Palace. This Victorian view of the Alexandra Palace terminus with GNR 0-4-4T No.697 in the foreground, shows the newly laid out streets and housing development in Muswell Hill which the railway had done so much to stimulate. Opened in 1873, the branch preserved a countrified atmosphere, but plans before and after World War 2 to electrify the line were never carried through. Bus and tram competition became severe and the poorly patronised and run down push-pull services to gaslit stations finally came to an end on 5 July 1954.
Photo: Alexandra Palace Arts Society

ESSEX LIGHT RAILWAYS OF THE GER

ELSENHAM & THAXTED. In Bradshaw's Guide the time table heading for this 5½ mile branch of the LNER was styled: ELSENHAM & THAXTED (Light Railway) – as if to infer that something unexpected might happen on the journey to Thaxted. The thing that upset the ordinary non-enthusiast traveller on light railways was that passengers and potatoes were not treated as separate commodities. Your train was liable to be shunted into a siding, carriages and all, to deposit a van of layer's mash or to collect a coal wagon. This is what is about to happen at Sibley's on the Elsenham & Thaxted Light Railway with J69 0-6-0T No.68579 in charge on 15 September 1952, just a few days before the passenger service ended. This line, which left the GE Cambridge line at Elsenham, was a late promotion of the GER, being opened in 1913. Its carriages began life on ambulance trains in World War 1; converted for conductor guard use, they were fitted with steps for boarding at the halts. *Photo: Derek Chaplin*

KELVEDON & TOLLESBURY. The premier LNER branch for sheer quaintness was the Kelvedon & Tollesbury. It opened in 1904 from Kelvedon on the main Colchester line, to Tollesbury, and its mixed trains of GE six wheelers, vans and wagons rattled over ungated crossings with American style cattle guards. The quaintness increased after 1928 when the low-slung tram-type carriages with end platforms arrived on the line from the Wisbech & Upwell Tramway. A GE J67 0-6-0T worked the branch, and a lively time was assured to travellers on the coach balcony as the loco tackled the long 1 in 50 banks. The trains were

too slow for the local population who had long since gone by bus. The Wilkin's Jam siding kept the line active for freight as far as Tiptree until complete closure in 1962 but the passenger trains disappeared in 1951.

ABOVE: No.68616, a J67 0-6-0T, gets ready to leave Tollesbury with the 12.50pm for Kelvedon on 19 September 1950. *Photo: Ian L. Wright*

RIGHT: One of the gaslit coaches, No.60461, was bought by a film company and figured in 'The Titfield Thunderbolt' film. The other Wisbech & Upwell coach, No.60462, was restored for preservation but was later broken up. Fortunately the body of No.60461 survived and is undergoing preservation at the Rutland Railway Museum, Cottesmore.
Photo: Ian L. Wright

EAST SUFFOLK GOODS

A pick up goods in Suffolk: GE J15 0-6-0 No.65467 sets out from Woodbridge on the East Suffolk line with the Ipswich to Framlington pick up goods on 10 October 1956. Familiar on Britain's railways for so long, the slow moving loose coupled goods train had become an uneconomic mode of freight handling by the 1950s. Following the Beeching Report hundreds of country goods yards were closed and ruthlessly stripped of their sidings.
Photo: R. C. Riley

SUFFOLK DOCK LINES

ABOVE: The LNER was an enthusiastic user of Sentinel steam shunters in the 1920s and 30s and a number were put to work in Permanent Way Department sleeper depots. The makers claimed that they used less than half the coal burned by ordinary shunting locos on the same work. In this picture taken in 1951, a flagman escorts LNER Y1 Sentinel 0-4-0T No.8130 safely through the streets of Lowestoft after shunting duties along the quays. This engine was the first LNER Sentinel and spent its entire working life at Lowestoft from 1925 to 1956.
Photo: O. H. Prosser

LEFT: If you had not seen one of these engines before, you could be forgiven for thinking this was a steam powered brake van! LNER J70 No.7135 was in fact a six-coupled tram loco from the Wisbech & Upwell Tramway in Cambridgeshire, one of twelve built by the GER between 1903 and 1921. With the statutory Board of Trade skirt removed, No.7135 makes its cowcatchered way over the Ipswich Dock lines in LNER days. Their initial allocation which did not greatly vary over the years was three to Yarmouth and six to Ipswich for dock work, the remaining three to Wisbech. *Photo: V. R. Webster*

MID SUFFOLK LIGHT RAILWAY

This delightfully rural and remote light railway began at Haughley Junction where the Great Eastern's Ipswich-Norwich and Ipswich-Bury St Edmunds lines diverged. It had an independent existence until after the grouping in 1923 and ancient GE gaslit six-wheelers were used in the mixed trains until more modern GE bogie vehicles were put on about 1951. Much youthful discomfort might be put up with in visiting lines like this. The author remembers, wrapped in a tent, sleeping in Haughley Junction waiting room so as to be ready for the morning mixed train to Laxfield.

BELOW: The ubiquitous J15 0-6-0 was the usual motive power. In this picture GE J15 No.65447 pauses at Kenton with the morning mixed train from Haughley to Laxfield on 29 June 1952.

ABOVE: One of the original Mid Suffolk nameboards at Aspall. *Photos: Ian L. Wright*

Deep in rural Suffolk, the fireman of the J15 opens the gate and sees the Laxfield mixed train across a country lane near Mendlesham, 29 June 1952.
Photo: Ian L. Wright

NORFOLK LINES GER

ABOVE: GE D16/3 4-4-0 No.62577 takes the sharp curve away from Wells on Sea with an afternoon train to Heacham, Easter 1952. The seemingly curious siting of the signal post is explained by the 180 degree curve round which trains approached Wells Junction.
Photo: P. B. Whitehouse MHL

RIGHT: Approaching Thetford Bridge station on the Thetford-Bury St Edmunds branch line. GE F6 2-4-2T No.67237 is working the 8.15am from Thetford to Bury on 25 May 1951. Since its last passenger trains in 1953 Thetford Bridge station has been a youth hostel and a Council road depot.
Photo: P. M. Alexander MHL

RIGHT: The GE's 18 mile Wells on Sea-Heacham line was a link between the Hunstanton and the Wells-Dereham branches "across the top of Norfolk". It was a line of wide open spaces and the 4-4-0s ran fast from village to village, bending and swaying the waves of ripening corn. The guard's flag is raised as D16/3 No.62557 moves the 1.35pm Wells to Heacham out of Burnham Market on 17 May 1952. Two weeks later the passenger trains had gone.
Photo: Ian L. Wright

BELOW: The seaside terminus at Hunstanton GER in Victorian times. GE "Intermediate" 2-4-0 No.479 waits to leave for Kings Lynn.
Photo: Lens of Sutton

CAMBRIDGE E4s

These well loved GE 2-4-0s were a familiar sight on branch line work throughout East Anglia. J. Holden built 100 of them between 1891 and 1902 but only 18 of the class survived to carry BR numbers. Following a policy of swopping locos about the system, the LNER sent six to the North East in 1935 where they worked on the Darlington-Penrith and Middleton-in-Teesdale branches, returning to the GE Section six years later. The Mildenhall, Waveney Valley and Stour Valley branches were happy hunting grounds of the E4 type in the 1950s.

TOP: Mildenhall terminus on 19 September 1953 with E4 2-4-0 No.62785 waiting to leave for Cambridge with a two coach train. This was the last survivor of the class. Withdrawn in 1959, No.62785 was restored to GER condition for preservation, originally at Clapham, as GER No.490. In the summer of 1985 it was sent from the NRM, York, to Norfolk on loan to Alan Bloom's railway museum at Bressingham. *Photo: Ian L. Wright*

BOTTOM: Cambridge E4 No.62783 runs on to the hand operated turntable at Mildenhall on 15 April 1952. *Photo: P. B. Whitehouse MHL*

OPPOSITE:
Country branch par excellence was the GE Stour Valley line between Cambridge and Marks Tey and its associated branch from Long Melford to Bury St Edmunds. Until the late 1950s elderly E4s and J15s still headed passenger trains through meadows and river scenery made famous by the painter John Constable. Here, through the carriage window, were the great wool churches of Lavenham and Long Melford and the pretty villages of Cavendish and Clare. If you were lucky you might see watercolour pictures of some of them in your Gresley compartment for it was the LNER's habit to promote its side of England with pictures of the places that the railway served. In

this photograph taken from a Bury-Colchester train, the Sudbury signalman stands ready to hand the staff to the driver of E4 No.62797 on a July day in 1957. *Photo: Ian L. Wright*

M & GN CROMER BRANCH

LEFT: D16/3 4-4-0 No.62533 pulls away from West Runton with a train from Melton Constable to Cromer Beach on 3 September 1955. *Photo: P. H. Groom*

NORFOLK & SUFFOLK JOINT RAILWAY

BELOW: A touch of Edwardian opulence still lingers at Mundesley on Sea station on the GE and M&GN Joint line from North Walsham to Cromer Beach on 28 August 1964 as a DMU waits to leave on the 11.51am service to North Walsham. Mundesley became a terminus when this coastal route to Cromer Beach was closed beyond Mundesley in 1953. The branch closed down on 5 October 1964 and, sadly, so did the camping coach holidays (left of picture) which the LNER had introduced at selected places in 1933. *Photo: Ian L. Wright*

M & GN JOINT RAILWAY

The Midland & Great Northern Joint Railway was almost the only competition experienced by the GER in East Anglia. M&GN lines ran from west of Bourne in Lincolnshire through Sutton Bridge and Fakenham to Yarmouth and much of the route was single line. From its country HQ at Melton Constable, deep in Norfolk fields, the M&GN worked branches to Norwich and Cromer. In 1936 the LNER took over responsibility for the M&GN's locomotives, a curious mixture of GN and Midland types, although some were of distinctive M&GN origin. Almost the whole railway closed in 1959 but BR still operates between Cromer and Sheringham.

ABOVE: In 1900 a stud of forty 4-4-0s of S.W. Johnson's design were working express passenger trains on the M&GN Joint Railway. Sharp Stewart built thirty-three of them between 1894 and 1896 – handsomely curved engines with 6′6″ driving wheels – and a further seven were completed by Beyer Peacock in 1899. LNER No.076, of the 1899 series, brings a touch of Midland Railway elegance to the GNR's Peterborough North station in August 1939 as it waits to pick up through coaches from King's Cross for the M&GN Peterborough branch and hurry them across the Fens to Sutton Bridge and on by the long way round to Cromer. No.076 carries the LNER livery adopted for M&GN engines in 1936. The "0" was the LNER's way of avoiding duplication with its own locomotive numbering scheme.
Photo: Dr Ian C. Allen

RIGHT: The Sheringham East starter is pulled off for a Cromer train in 1963 and the signalman leaves the box to hand over the token. GN somersault signals lasted a long time on the M&GN, whose atmosphere lives on in the North Norfolk Railway at Sheringham.
Photo: Ian L. Wright

DUNSTABLE BRANCH GNR

The northern edge of Dunstable's chalk escarpment in Bedfordshire might seem an unlikely venue for LNW and GNR trains but Dunstable North station on the LNW was the exchange point for passengers from Leighton Buzzard who joined the Great Northern train for Luton (Bute Street), Welwyn and Hatfield. It was a useful cross country route and a pretty run, particularly along the valley of the infant River Lea and through the woods near Ayot.

GNR services opened between Welwyn and Luton on 1 September 1860 though the Dunstable-Luton part of the branch was fully opened on 3 May 1858 and was worked for two years by LNW trains. Until the end of steam on the branch GE Class N7 0-6-2 tanks worked the passenger services, the heavier GNR N2 tanks having been taken off because of weight restrictions.

The Dunstable-Hatfield trains were discontinued from 26 April 1965 and the freight traffic ended on 3 January 1966 but a connection with the Midland main line at Luton, put in during 1966, allows access to an oil terminal at Dunstable.

RIGHT: GE Class N7 0-6-2T No.69618 leaves Luton Hoo on the 6.25pm Dunstable to Hatfield train on 10 July 1959. *Photo: J. F. Aylard*

THE WISBECH & UPWELL TRAMWAY

LEFT: Under the terms of the 1870 Tramways Act the GER promoted the unique Wisbech & Upwell Tramway to serve the fertile fenland potato fields, orchards and village communities along the course of the Wisbech Canal. A passenger service was operating throughout in 1884 using 0-4-0 and 0-6-0 tram engines built by the GER for working this roadside line. Board of Trade regulations required cowcatchers, enclosed motion and condensers to be fitted to the engines. J70 0-6-0 tram engine No.68225 is seen working on the tramway in early BR days.
Photo: H. Coates & Sons, Wisbech

IN RURAL BUCKS

ABOVE: At one time the Metropolitan Railway extended far into the Buckinghamshire countryside beyond Aylesbury and joined the LNWR at Verney Junction, fifty miles from its London Baker Street headquarters. In 1899 the Great Central main line used the route from Quainton Road to Marylebone, and the Verney line, now a branch, was put under the management of the Met&GC Joint Committee from 1906. In this picture a motor-fitted GE F7 2-4-2 No.8307 is making a stop at the miles-from-nowhere station of Winslow Road, on its way to Verney Junction on 2 May 1936. The loco is propelling an ex GC motor-fitted twelve wheeler.
Photo: H. C. Casserley MHL

RIGHT: Chesham was once the outer terminus of the Met, deep in leafy Buckinghamshire. Then, when the Met was extended to Aylesbury and beyond, a route was chosen through Amersham and Great Missenden, and Chesham was left on a branch served from Chalfont & Latimer. The LNER took over the steam operation of the Metropolitan line in the 1930s and the Chesham branch was operated by push-pull trains from 1941. Here is a memory of the Chesham branch steam days, a post nationalisation picture of Robinson GC 4-4-2T No.67418 hurrying the ancient "Chesham set" of carriages on their way to Chalfont & Latimer before the branch was electrified in 1961. One of these carriages is preserved in the London Transport Museum, while others require extensive restoration on the Bluebell Railway. *Photo: Peter Hay*

IN WALES AND WIRRAL

ABOVE: The Great Central gained a foothold in Wales due to its absorption of the Wrexham, Mold & Connah's Quay Railway, a local line with interests in exploiting the Flintshire coalfield. The GC also had its own line in Wirral. Wrexham Exchange station is the setting on 9 May 1949 as Parker GC N5 0-6-2T No.9289 comes in on a train from New Brighton and Connah's Quay. *Photo: P. M. Alexander MHL*

PEAK DISTRICT BRANCH

RIGHT: Pennine terminus: GC C14 4-4-2T No.67451 is just inside the borders of Derbyshire's Peak District and waiting among sombre hills at Hayfield to take out the 3.18pm to Manchester London Road on 26 August 1956. The Hayfield branch, a GC & Midland Joint line from a junction at New Mills, was closed to passengers from 5 January 1970.
Photo: T. J. Edgington

GNR IN NOTTS, DERBY & STAFFS

The GNR sent out a branch from Grantham to reach Nottingham as early as 1857, but an extension westwards around Nottingham in 1875 broke the Midland Railway's monopoly of the coalfield and gave the GN access to Derby and Burton. By running powers over the North Stafford, the GN later reached Stafford over its own isolated Stafford & Uttoxeter line.

RIGHT: A GN J6 0-6-0, No.3623, heads away from Stafford "wrong line running" to gain the GN Stafford & Uttoxeter line with a Grantham train in 1933. Outside Stafford a chilling notice warned GNR drivers – "WHISTLE AT CEMETERY CROSSING". Passenger traffic ceased in 1939. *Photo: P. S. Kendrick*

BELOW: Happier days at late-lamented Nottingham Victoria from which every vestige of the railway has today been swept away. An LNER branch train is ready to leave with GN D2 4-4-0 No.2194 on a turn to Grantham in July 1947, twenty years before closure. *Photo: H. C. Casserley MHL*

THE GN & LNW EAST LEICESTERSHIRE JOINT LINE

Far to the east of Leicester the cold empty uplands around Tilton are crossed by deep cuttings and massive brick viaducts bestride the valleys. These are the remains of the GN&LNW Joint line (opened 1879) which ran from Welham Junction on the LNW near Market Harborough, to Melton Mowbray and Stathern Junction where it divided for Nottingham and for Grantham. A lordly waiting room was provided at Redmile for the Duke of Rutland's hunting guests at Belvoir Castle and Melton Mowbray station was laid out on a lavish scale.

The GN built its own line to Leicester from a junction near John o'Gaunt and opened its Belgrave Road station in 1883 with passenger services to Grantham, Newark and Peterborough. By 1953 the train illustrated here was all that was left — the 1.0pm SO Leicester Belgrave Road to Melton Mowbray North and back. At summer weekends the picture totally changed. Hundreds of holiday makers headed out of Belgrave Road aboard seaside specials for Skegness and Mablethorpe. Hauled by GN J6 0-6-0s and later by Thompson B1 4-6-0s, these trains carried on running until 9 September 1962.

OPPOSITE: Melton Mowbray North, on the GN&LNW Joint line, from which the glory had long departed when this photograph was taken on 7 November 1953. GC J11 0-6-0 No.64438 stands by glassless iron canopies waiting to depart with the 1.55pm SO for Leicester Belgrave Road. One day in 1954 when this train was derailed in the Melton sidings, the dozen or so regulars arrived in Leicester by brake van!
Photo: Ian L. Wright

LEICESTER GNR BELGRAVE ROAD

TOP: No.64438 under the iron arches of Leicester Belgrave Road station with the 1.55pm return Saturday working from Melton Mowbray North on the same day. At this time the stored Skegness excursion stock shared the station with British Road Services. A workmen's service continued to be provided between Belgrave Road and John o'Gaunt until 29 April 1957.
Photo: Ian L. Wright

BOTTOM: Great Northern period piece: a memory of the 1930s as GN D2 4-4-0 No.3045 waits at Leicester Belgrave Road with a good head of steam ready to take out a Grantham train on 29 June 1934.
Photo: V. R. Webster

THE GNR IN LINCOLNSHIRE

ABOVE: Local trains at Lincoln LNER, around 1945. The GN D2 4-4-0s are Nos.4332 and 4328. The tender of No.4328 is painted in wartime "NE" lettering. *Photo: Lens of Sutton*

RIGHT: Horncastle branch train at Woodhall Junction on the Lincoln-Boston line in 1953. The loco is a GC A5 4-6-2T, No.69804, weighing 85¾ tons. The coaching stock, an articulated set converted from GN steam railmotors, weighs only 37 tons! Lines in Lincolnshire and East Anglia were expensive for the LNER to run because of the enormous number of manned level crossings. Horncastle lost its passenger trains in 1954 and the site of the line is now a seven mile stretch of the Viking Trail across Lincolnshire. *Photo: O. H. Prosser*

THE MARQUIS OF EXETER'S RAILWAYS

The Midland was the first railway in Stamford, a beautiful stone town on the Jurassic limestone belt, and much under the domination of the Marquis of Exeter (Lord Burghley of Burghley House). The Marquis built two branch lines from an attractive little terminal station beside the Welland (Stamford East); one to Essendine on the GNR East coast main line (opened in 1856) and one to Wansford on the LNWR (1867), both branches being worked by the GNR.

TOP: The Stamford branch was one of those lines where very little ever seemed to change. Engines were coaled by primitive bucket and crane and GN tank engines lived out their long declining years. GN C12 4-4-2T No.67365 is shunting the yard at Stamford East in 1953. The pile of old rope on the buffer beam strikes a note of typical branch line informality!

To the left of the picture is the MR Peterborough-Leicester line and the GN connection between the two railways. On 4 March 1957 the Essendine trains were diverted over this line into Stamford Town and this allowed the East station to be closed, although freight traffic continued for another six years. *Photo: P. M. Alexander MHL*

BOTTOM: The GN C12 4-4-2 tanks were aristocratic looking engines, accounted for maybe by the Atlantic wheel arrangement, the expansive front end and the set back smokebox and boiler. This classic GN branch train with No.4508 in charge, was photographed at Essendine in LNER days. Informality again as a sack of merchandise travels to Stamford on the front of the engine! *Photo: Dr Ian C. Allen*

SOUTH HUMBERSIDE GCR

RIGHT: The Manchester, Sheffield & Lincolnshire (later GCR) spent a great deal of money developing seaside amenities at Cleethorpes to which it had opened a branch from Grimsby Docks in 1863. Accommodation for excursion traffic, particularly from industrial Yorkshire, demanded an expansion to six platforms. Some interesting antique local trains were also run out of Cleethorpes by the LNER in pre war days. MS&L 4-4-0 No.5684, in immaculate condition, prepares to leave Cleethorpes with a train of six wheelers for New Holland in LNER days. *Photo: Lens of Sutton*

BELOW: "The LNER also operated trams". The Grimsby & Immingham Electric Tramway was opened by the GCR in 1912 to provide transport for Immingham dockworkers. Seven miles long, the line started at Corporation Bridge, Grimsby, and ran the first mile in the public street and then into open country. The GCR's single deck bogie cars had a central area for milk and merchandise. It was a line of great character, but closed down in favour of buses on 1 July 1961. Car No.16 stands at the Tramway Station in Corporation Road, Grimsby, on 23 May 1953. Car No.14 is in the care of the National Tramway Museum but is in store awaiting restoration. *Photo: O. H. Prosser*

LEFT: While the Humber remained unbridged, British Rail's railhead pier and ferry at New Holland continued to be important for South Humberside's access to Hull, both for passengers and cars, and at one time live cattle as well! The Humber Bridge finally opened for road traffic in 1981 and so ended the New Holland Pier trains and the well-loved Sealink ferry *Farringford* with its memories of the Isle of Wight. A memory of 1946 is this scene at New Holland Town with Parker GC N5 0-6-2T No.5898 ready to set off for Immingham.
Photo: H. C. Casserley MHL

HORNSEA BRANCH NER

BELOW: North of the Humber two North Eastern Railway branches left Hull and ran to the North Sea coast resorts of Hornsea and Withernsea. Both were subjected to savage cuts in services and then axed by Dr Beeching in 1964. Hull will remember happier days like this one, 2 August 1949, as NE A6 4-6-2T No.69795 brings a train into the excursion platform at Hornsea for a day at the seaside.
Photo: T. J. Edgington

WEST YORKSHIRE

LEFT: A typical NE branch scene near the wooded north end of Bramhope Tunnel on 19 September 1941 with NE G5 0-4-4T No.1911 at the head of a Leeds-Ilkley train.
Photo: M. N. Clay/Lens of Sutton

NORTH YORKSHIRE

BELOW: The NER branch to Richmond was worked by trains from Darlington which took to the Richmond line after a run down the East Coast main line to Eryholme Junction. Although the trains have gone, Richmond's attractive gabled station by the Yorkshire architect G. T. Andrews remains. Refurbished by Richmondshire District Council, it has been leased as a garden centre — a rare example of imagination and enterprise that could be widely followed. In steam days back in August 1950 LNER A5 4-6-2T No.69840 stands at Richmond with a Darlington train.
Photo: P. B. Whitehouse MHL

WHITBY-SCARBOROUGH

ABOVE: Whitby Town on 3 May 1958. One of the handsome NE A8 4-6-2 tanks, No.69861 leads the 4.20pm Middlesbrough-Whitby-Scarborough in reverse out of Whitby Town for the climb to West Cliff. Out of the picture (left) is the "real" train engine, BR standard 2-6-4T No.80116 which is banking at the rear. No.80116 will continue its journey forward to Scarborough after another reversal at West Cliff. The 4.20pm ex Middlesbrough was the last through train from Middlesbrough to Whitby by the coast route via Loftus on 3 May 1958. The line closed, BR said, because substantial expenditure was necessary on the viaducts and tunnels. *Photo: Ian L. Wright*

RIGHT: Resting at Scarborough: LNER No.689, a NE A6 4-6-2T originally built as a 4-6-0 tank for the steeply graded Whitby-Scarborough route, takes a spell from duty in the summer of 1937.
Photo: N. Shepherd/R. C. Riley Collection

IN WENSLEYDALE

Travellers on the Settle & Carlisle line may have looked out of the window at Garsdale and wondered at the narrow ribbon of trackless railway winding away eastwards on a ledge of steep mountainside. The line was the Midland's branch from Garsdale (one time Hawes Junction) to Hawes, and LNER trains worked over it on the last leg of the 39¾ mile run from Northallerton to Garsdale. This long and very beautiful branch was closed to passengers between Northallerton and Hawes on 26 April 1954, the ex MR portion on to Garsdale lasting another five years.

TOP: The signalman exchanges the staff with the crew of a Northallerton-Garsdale train at Leeming Bar on the Wensleydale Branch in March 1954. The capstan in the signal box is for manual operation of Leeming Bar level crossing gates where the line is crossed by the Roman Watling Street.
Photo: W. T. Baldwin

BOTTOM: Until it was lost to road transport, milk was an important traffic on the branch, first in churns by six wheel milk vans, and later by rail tank wagons. The old NER's class M1 4-4-0s involved in the 1895 Race to the North, were sometimes put on this duty and here is one of them, LNER D17/1 No.1638, standing at Hawes with a Wensleydale milk train in June 1932. Hawes station remains intact as a North Yorkshire National Park Information Centre and Museum. Although jointly used by the MR and NER it is a typical MR building. *Photo: Dr Ian C. Allen*

ABOVE: Another veteran of the milk trains – NE F8 2-4-2T No.40 at Leyburn with milk vans in 1935, and a good opportunity to study the style of LNER locomotive lettering and passenger stock numerals of the period. *Photo: H. C. Casserley*

A row blew up in the peaceful valley only one week after the closure to passengers. As a national newspaper put it: "ONE HORSE SPECIAL ANNOYS THE TRAINLESS!" BR put on a horse box special to Leyburn to take a racehorse to Northallerton just days after the passenger trains finished, and 6000 locals were said to be outraged. BR's answer was that this train was full even though it contained only one horse. It also paid its way. Parcels, newspaper, fish and milk trains would continue to be run on the branch line, the spokesman said. Leyburn lost its passenger trains because they were not full!

OVER STAINMORE

The South Durham & Lancashire Union Railway opened its line across the Pennines between Barnard Castle, Kirkby Stephen and Tebay in 1861. It was a line of severe gradients across wild country and the well known summit of 1,370 feet over Stainmore was the highest point on a through passenger route on an English railway. Built to carry Durham coal and coke to the Barrow iron furnaces and to move iron ore in the opposite direction this NER line had some outstanding iron viaducts at Deepdale and Belah, both built under the direction of Sir Thomas Bouch, the engineer to the ill-fated Tay Bridge. For years the LNER allowed nothing heavier than J21 Class 0-6-0s over these viaducts. Belah, on the Cumbrian side of Stainmore, was 1,040 feet long, 192 feet high and had 16 spans. Crossing the Belah girders with head out of the carriage window was almost an air travel experience, breathtaking in its drama. It is sad to think it is unrepeatable. Piece by piece the viaduct's spindly structure was dismantled soon after the last train ran on 20 January 1962.

BR Standard 2-6-0s Nos.78016 and 76024 double head the 11.20 (SO) Blackpool North-Newcastle train over Belah Viaduct on 18 July 1959.
Photo: Robert Leslie

OPPOSITE, TOP: On 8 August 1950 one of the hardworking mixed traffic NE Class J21 0-6-0s, No.65038, calls at Ravenstonedale on the Tebay line with a two coach Darlington train.
Photo: P. B. Whitehouse MHL

OPPOSITE, BOTTOM: Working in tandem on 'The Stainmore Limited', a special last train from Darlington over the NER Penrith and Tebay lines, BR Class 3 and 4 2-6-0s Nos.77003 and 76049 halt in winter sunshine at Kirkby Stephen East on 20 January 1962.
Photo: Ian L. Wright

RAVENSTONEDAL
Nº 65038

77003
X76

ALSTON BRANCH

Alston, 1000 feet up in the Cumbrian Pennines, is England's highest market town. It can also be a hostile place when in the grip of winter. The railway came to Alston as a steeply graded single line branch from Haltwhistle on the NER Newcastle-Carlisle line and it followed the narrowing valley of the South Tyne river. The LNER considered closure to passengers in 1929 but the roads to Haltwhistle were too poor for replacement buses. Alston, which is just in Cumbria, had cause to be thankful for the railway when winter snow cut the town off from the outside world. NE J21s and G5s did good work on the branch. Modern locos too, like BR Class 3MT 2-6-0 No.77011 which worked passenger turns after transfer to Alston in 1955. Alston had that appealing branch line feature, the one road engine shed, and its all-over station roof gave some protection against the elements.

TOP: This is the setting for LNER J39 0-6-0 No.64858, a Hull Dairycoates engine, whose driver picks his way across the track after bringing in a train from Haltwhistle in March 1954.

In Alston the old ways are gone, but a 2ft gauge venture, the South Tynedale Railway, is extending a passenger line along the trackbed.

Photo: O. H. Prosser

BOTTOM: Lambley, a picturesque wayside station on the Alston branch, sited where a stone viaduct took the railway over the South Tyne.

Photo: Lens of Sutton

TYNE & WEAR

Sir Vincent Raven's NER 0-8-0s of LNER Class Q6 were a sound, sturdy design, simple in outline and well suited to their task of moving heavy mineral and freight trains over the steeply graded branches of the NER's industrial heartland. The Q6 class was so successful that 120 were built between 1913 and 1921 and they all remained at work for more than a decade after Nationalisation. You would find them moving coal from Durham pits to Hartlepool Docks or to power stations, or battling eastwards through How Mill on the Carlisle Road with mineral trains for ICI at Prudhoe. They were at home moving steel around Middlesbrough, sometimes banking or piloting, sometimes working in pairs. Happily a Q6 has been preserved on the North Yorkshire Moors Railway as LNER No.3395, though its role as a passenger performer takes some getting used to!

Here, NER Class Q6 0-8-0 No.63436 storms away from South Pelaw with a coal train for the Consett area on 24 August 1966. *Photo: Ian S. Carr*

LANGHOLM NBR

Langholm branch trains worked to and from Carlisle with a 14 mile run over the Waverley Route to Riddings, all on the English side of the border. This was followed by a further 7 miles on the single line branch between Riddings and Langholm, all of it in Scotland once you had crossed the viaduct over the Liddel Water.

The regular branch engine was Ivatt Class 4 No.43139 which came new to Carlisle Canal shed in 1951 and spent much of its working life on the line. During 1962-63 Fowler 2-6-4 tanks borrowed from Kingmoor shed were regularly working the branch and contributing an equally alien atmosphere of 'Midlandisation' to this North British byway. The line opened from Riddings to Langholm on 18 April 1864 and lasted just over a century, closing to passengers on 15 June 1964 and finally to freight on 18 September 1967. The main engineering features of the branch were the high viaducts at Riddings and Gilnockie.

In this view, an LNER flavour temporarily returns to Langholm as J39 0-6-0 No.64964 sets out with the 3.23pm to Carlisle on 16 April 1960.
Photo: Robert Leslie

LNER STEAM RAILCARS IN SCOTLAND AND THE BORDER

The LNER was an enthusiastic user of steam railcars and many a branch line stayed alive in the 1930s because of their economical working. Between 1924 and 1930 around 90 steam railcars were put in traffic. The three examples illustrated here are working Scottish branch lines but the cars were to be seen all over the LNER, particularly in the North East. The first Sentinel cars were of the 2 cylinder type. Later 6 and 12 cylinder types appeared.

It was a nice touch when Gresley decided to name them. His Pacifics got the names of racehorses, sometimes with bizarre results. The jaunty names of the railcars came from those previously carried by horse drawn stage or mail coaches in Britain. *Bang Up* was a Clayton car that worked out of Lowestoft. Sentinel car *Tally Ho* reduced running costs on the Brough-Hull service. *Hope* and *Courier* ran coupled together out of Whitby, and *Tantivy* had a summer season at Yarmouth on the M&GN with disastrous results. While serving the brand new halts out to Potter Heigham it set fire to every grass bank within sight. Not surprisingly, *Tantivy* was never seen in Yarmouth again!

BELOW: *Quicksilver* was a 6 cylinder railcar working in Scotland. It was also the name of an A4 streamlined Pacific, and in 1936 a Railway Magazine correspondent noted the two *Quicksilvers* together at Edinburgh Waverley. *Photo: E. R. Wethersett P. B. Whitehouse Collection*

NORTH BRITISH BRANCHES

LEFT: At one time the NBR promoted the Aberfoyle branch as its route to the Trossachs, rivalling the Caledonian's Callander way. Certainly the Aberfoyle branch was outstanding for its scenery and Glasgow folk took the train to the falls at Campsie Glen before road coaches tempted them away. The branch left the Glasgow-Edinburgh main line at Lenzie and served Kirkintilloch, Lennoxtown and Blanefield. Beyond Blanefield the service was more sparse, and No.312 *Retaliator,* a 100 HP 6 cylinder railcar, is recorded amid a backdrop of mountains at Aberfoyle in December 1932, ready to leave on the Blanefield run. Passenger traffic ceased in 1951. *Photo: Dr Ian C. Allen*

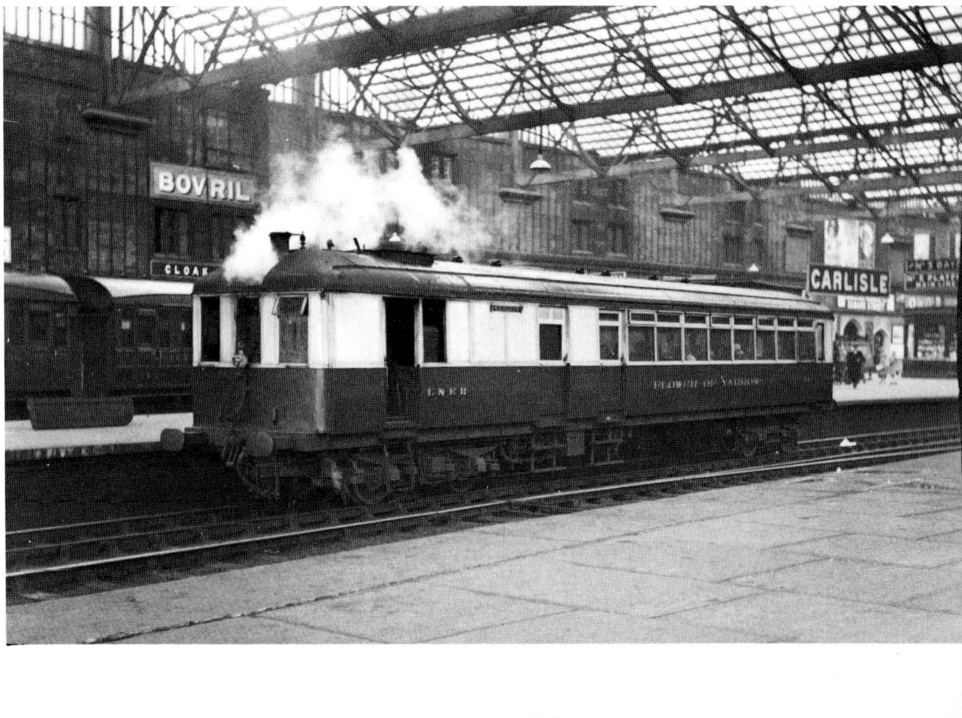

RIGHT: Except for the Sentinel railcar, the setting is not much changed – Carlisle Citadel east side. 6 cylinder Sentinel No.31 *Flower of Yarrow,* wearing the green and cream livery that was usual with these cars, waits with steam up for a trip on the Langholm branch on 25 August 1935, a run that will take it over the Waverley Route to Riddings and into Scotland. This branch survived until 1964 and the Waverley line only outlived it by five years.

Steam railcars were run with a driver and fireman but no guard. Ultimately they proved unsuccessful. The light units did not offer enough accommodation in peak hours and to put them on at slack times only, was uneconomic. Within a few years after World War 2 they had all gone. *Photo: P. S. Kendrick*

NBR
IN THE BORDER COUNTIES

From St Boswells on the Waverley Route a branch line ran to Kelso and on to Tweedmouth, and until the middle 1950s you might find a NB C15 Atlantic tank at the head of your train which ambled through the lonely agricultural lands beside the Tweed. Through Roxburgh and Kelso your train was in Scotland and on the NB. Over the Border to Coldstream you were in Northumberland and on the NE. It was too unpromising a land to support a railway in the climate of the Beeching Report and perhaps the townsfolk of Kelso and Coldstream did not notice much when their two trains a day were discontinued from 15 June 1964.

Down to one coach, a Gresley brake-composite, the Tweedmouth-St Boswells branch train is seen leaving Kelso in 1962. The loco is BR Standard Class 2MT No.78047. *Photo: Michael Mensing*

Polmont, on the Glasgow-Edinburgh main line, was the junction for the local service to Falkirk (Grahamston) and this well judged study of a Scottish branch train in the landscape depicts NB C16 4-4-2T No.67494 curving away from the main line at Polmont with the late afternoon two coach stopping train for Falkirk (Grahamston) and Grangemouth in the summer of 1958. *Photo: D. A. Anderson*

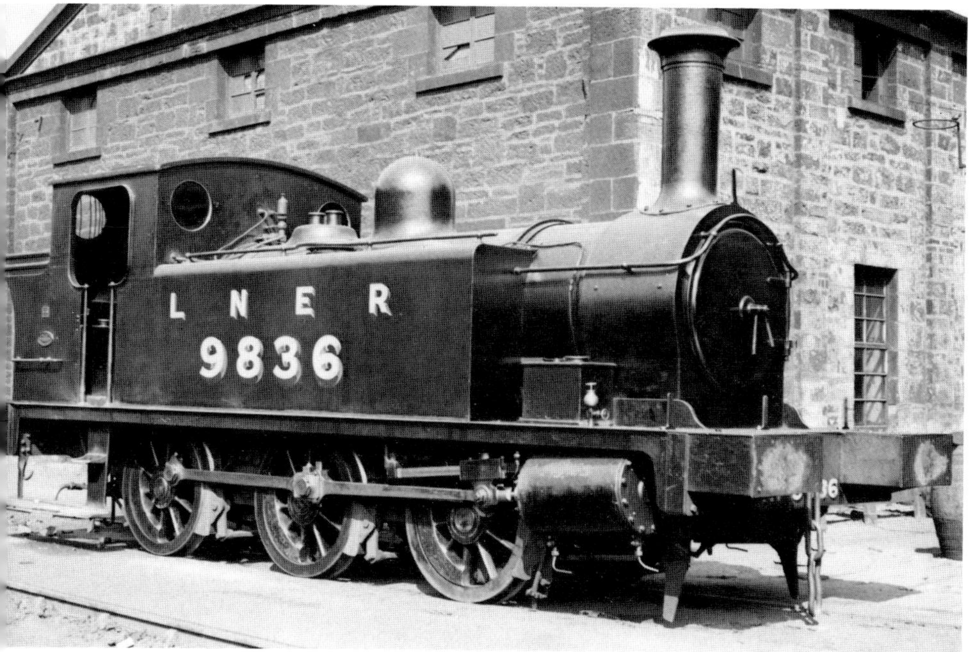

MORE NBR BRANCHES

LEFT: Leith Dock shunter. This most winsome of Scottish tank engines is a NB J88 posing in ex works condition among the crumbling masonry of Edinburgh St Margaret's shed in 1937. St Margaret's, which must have held the prize for the worst smoke pollution in Britain, certainly kept its J88s looking superb. No.9836 was one of several St Margaret's engines allocated to working over the sharp curves of the Leith Dock lines as may be guessed by the short wheelbase and wood buffers. Even in black and white photography this engine well illustrates the beauty of LNER shaded letters and numerals.
Photo: L. Hanson

INVERBERVIE BRANCH

RIGHT: Michael Mensing's nicely composed picture encapsulates the relaxed atmosphere of the minor Scottish branch line. It also sums up a favourite experience of steam age childhoods. The six little figures on the footbridge, some with heads through the lattice, are waiting for the J37 to puff some steam up under the footbridge floorboards. No.64608 is at Gourdon in 1964, on the branch from Montrose to Inverbervie. Passenger traffic ceased in 1951 and freight traffic followed in 1966.
Photo: Michael Mensing

ROAD TO MALLAIG

This scenically unsurpassed line, the North British Railway's West Highland line from Craigendoran to Fort William and on to Mallaig, has always been difficult to operate. Weight restrictions and sharp curves on this heavily graded route have tested successive locomotive designers. After the formation of the LNER, Gresley sent 13 of his GNR Class K2 Moguls to take charge of the increasingly heavy passenger trains on the line. These West Highland 2-6-0s were named after Scottish lochs and they were fitted with protective side-windowed cabs against the severe northern climate.

TOP: LNER Class K2 No.4692 *Loch Eil* is being manually turned in the bleak shed yard at Mallaig, ready to head a passenger train to Fort William on 19 June 1937. At right is an ex-NBR Class J37 0-6-0. *Photo: H. C. Casserley*

BOTTOM: In 1937 Gresley produced the first of the K4s, No.3441 *Loch Long,* a 3 cylinder 2-6-0 designed especially for the West Highland which could take on a train of 300 tons single handed. One of these, resplendent in apple green, is No.3442 *The Great Marquis,* now preserved. Edward Thompson rebuilt K4 No.3445 *MacCailin Mor* at Doncaster in 1945, and this engine, designated Class K1/1, was the first of the numerous K1 Class mixed traffic engines produced from 1949 onwards by A. H. Peppercorn.

With the huge mass of Ben Nevis in the background, one of the West Highland K1s, No.62034, accelerates away from Banavie towards Corpach and the shores of Loch Eil with a Mallaig train, after a 10 mph 'slow' to cross the swing bridge over the Caledonian Canal. No.2005 of the K1 Class is preserved on the North Yorkshire Moors Railway. *Photo: Michael Mensing*

GNSR ST COMBS LIGHT RAILWAY

The Great North of Scotland Railway built only one line under the Light Railways Act of 1896, its light railway from Fraserburgh to St Combs, opened in 1903. Some trains ran mixed, and as the track was unfenced, the locos were fitted with the required cowcatchers, both fore and aft. Some pretty little Holmes D51 4-4-0 tanks were sent to work the line in the 1930s, to be followed by some not-so-pretty GE 2-4-2 tanks, among them 67157, the last F4. The latest engines to come were the Ivatt Class 2s, and No.46460 was cowcatcher-fitted like all the rest. Unfortunately for this photographer, Class 2 2-6-0 No.46464 was deputising for the regular engine on 25 August 1955 and no picture of a cowcatcher was possible. No.46464 now runs on the Strathspey Railway where it worked the first public steam hauled train on 16 July 1978.

In this 1955 picture at St Combs the fireman seems to be dematerialising, Dr Who style, in a cloud of steam! *Photo: Ian L. Wright*

GREAT NORTH OF SCOTLAND SPEYSIDE JUNCTION

RIGHT: The gently curved platforms and pastoral background of Craigellachie make a memorable setting for this vintage Speyside train, photographed on 2 August 1954. The engine, D41 4-4-0 No.62241 from Keith, and first coach are both of GNS parentage, with a Gresley brake third bringing up the rear. Craigellachie was the Great North's junction for the Elgin and Speyside lines.
Photo: P. B. Whitehouse MHL

KEITH

LEFT: Keith, like Elgin, was a Great North frontier post with the Highland, and relationships were none too cordial between the two companies. Even at the Grouping the two contestants found themselves in LNER or LMS camps. Pausing at Keith on a local train, GNS D38 4-4-0 No.6876, smartly turned out in LNER lined green and with a tender cab, poses for Mr Casserley's camera on 28 May 1930.
Photo: H. C. Casserley

ON SPEYSIDE

GNS D41 4-4-0 No.62241 again, calling at Ballindalloch with a Speyside train in October 1951. A 'home made' footbridge frames the train — an admirable piece of GNS economy that has been put together with lengths of old rail.
Photo: P. B. Whitehouse MHL

ABERDEEN: ON TRINITY QUAY

Road, rail and sea transport all meet in the granite streets of Aberdeen in 1955. GNS Manning Wardle Z4 0-4-2T No.68190 rests between shunting turns at Trinity Quay on the Aberdeen Harbour lines. There were four of these 0-4-2 tanks (GNS classes X and Y). This one was photographed on 26 August.

In classifying these unusual dock tanks the LNER finished up giving them the letter Z and this is perhaps as good a place as any to bring an end to this survey! *Photo: Ian L. Wright*